GEOMETRY
Workbook
Common Core Standards
Edition

Published by
TOPICAL REVIEW BOOK COMPANY
P. O. Box 328
Onsted, MI 49265-0328
www.topicalrbc.com

THE STATE EDUCATION DEPARTMENT / THE UNIVERSITY OF THE STATE OF NEW YORK / ALBANY, NY 12234

Common Core High School Math Reference Sheet
(Algebra I, Geometry, Algebra II)

CONVERSIONS

1 inch = 2.54 centimeters	1 kilometer = 0.62 mile	1 cup = 8 fluid ounces
1 meter = 39.37 inches	1 pound = 16 ounces	1 pint = 2 cups
1 mile = 5280 feet	1 pound = 0.454 kilograms	1 quart = 2 pints
1 mile = 1760 yards	1 kilogram = 2.2 pounds	1 gallon = 4 quarts
1 mile = 1.609 kilometers	1 ton = 2000 pounds	1 gallon = 3.785 liters
		1 liter = 0.264 gallon
		1 liter = 1000 cubic centimeters

FORMULAS

Triangle	$A = \dfrac{1}{2}bh$	Pythagorean Theorem	$a^2 + b^2 = c^2$
Parallelogram	$A = bh$	Quadratic Formula	$x = \dfrac{-b \pm \sqrt{b^2 - 4ac}}{2a}$
Circle	$A = \pi r^2$	Arithmetic Sequence	$a_n = a_1 + (n-1)d$
Circle	$C = \pi d$ or $C = 2\pi r$	Geometric Sequence	$a_n = a_1 r^{n-1}$
General Prisms	$V = Bh$	Geometric Series	$S_n = \dfrac{a_1 - a_1 r^n}{1 - r}$ where $r \neq 1$
Cylinder	$V = \pi r^2 h$	Radians	$1 \text{ radian} = \dfrac{180}{\pi} \text{ degrees}$
Sphere	$V = \dfrac{4}{3}\pi r^3$	Degrees	$1 \text{ degree} = \dfrac{\pi}{180} \text{ radians}$
Cone	$V = \dfrac{1}{3}\pi r^2 h$	Exponential Growth/Decay	$A = A_0 e^{k(t - t_0)} + B_0$
Pyramid	$V = \dfrac{1}{3}Bh$		

i

June 2017
Part I

Answer all 24 questions in this part. Each correct answer will receive 2 credits. Utilize the information provided for each question to determine your answer. Note that diagrams are not necessarily drawn to scale. For each statement or question, choose the word or expression that, of those given, best completes the statement or answers the question. [48]

1. In the diagram, $\triangle ABC \cong \triangle DEF$.

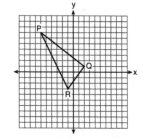

Which sequence of transformations maps $\triangle ABC$ onto $\triangle DEF$?
(1) a reflection over the x-axis followed by a translation
(2) a reflection over the y-axis followed by a translation
(3) a rotation of $180°$ about the origin followed by a translation
(4) a counterclockwise rotation of $90°$ about the origin followed by a translation

1 _____

2. On the set of axes, the vertices of $\triangle PQR$ have coordinates $P(-6, 7)$, $Q(2, 1)$, and $R(-1, -3)$. What is the area of $\triangle PQR$?
(1) 10
(2) 20
(3) 25
(4) 50

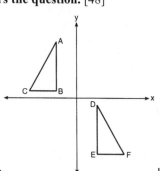

2 _____

3. In right triangle ABC, $m\angle C = 90°$. If $\cos B = \dfrac{5}{13}$, which function also equals $\dfrac{5}{13}$?

(1) $\tan A$ (2) $\tan B$ (3) $\sin A$ (4) $\sin B$

3 _____

4. In the diagram, $m\overset{\frown}{ABC} = 268°$. What is the number of degrees in the measure of $\angle ABC$?
(1) $134°$ (3) $68°$
(2) $92°$ (4) $46°$

4 _____

5. Given $\triangle MRO$ shown, with trapezoid $PTRO$, $MR = 9$, $MP = 2$, and $PO = 4$.

What is the length of $\overline{TR}$?
(1) 4.5 (3) 3
(2) 5 (4) 6

5 _____

6. A line segment is dilated by a scale factor of 2 centered at a point not on the line segment. Which statement regarding the relationship between the given line segment and its image is true?
(1) The line segments are perpendicular, and the image is one-half of the length of the given line segment.
(2) The line segments are perpendicular, and the image is twice the length of the given line segment.
(3) The line segments are parallel, and the image is twice the length of the given line segment.
(4) The line segments are parallel, and the image is one-half of the length of the given line segment.

6 _____

7. Which figure always has exactly four lines of reflection that map the figure onto itself?
(1) square (2) rectangle (3) regular octagon (4) equilateral triangle

7 _____

8. In the diagram of circle O, chord $\overline{DF}$ bisects chord $\overline{BC}$ at E.

If $BC = 12$ and FE is 5 more than DE, then FE is
(1) 13   (3) 6
(2) 9   (4) 4

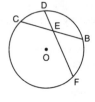

8 _____

9. Kelly is completing a proof based on the figure.

She was given that $\angle A \cong \angle EDF$, and has already proven $\overline{AB} \cong \overline{DE}$. Which pair of corresponding parts and triangle congruency method would *not* prove $\triangle ABC \cong \triangle DEF$?

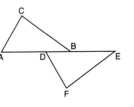

(1) $\overline{AC} \cong \overline{DF}$ and SAS   (3) $\angle C \cong \angle F$ and AAS
(2) $\overline{BC} \cong \overline{EF}$ and SAS   (4) $\angle CBA \cong \angle FED$ and ASA

9 _____

10. In the diagram, $\overline{DE}$ divides $\overline{AB}$ and $\overline{AC}$ proportionally, $m\angle C = 26°$, $m\angle A = 82°$, and $\overline{DF}$ bisects $\angle BDE$. The measure of angle DFB is
(1) 36°   (3) 72°
(2) 54°   (4) 82°

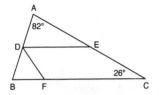

10 _____

11. Which set of statements would describe a parallelogram that can always be classified as a rhombus?
   I. Diagonals are perpendicular bisectors of each other.
   II. Diagonals bisect the angles from which they are drawn.
   III. Diagonals form four congruent isosceles right triangles.
(1) I and II   (2) I and III   (3) II and III   (4) I, II, and III

11 _____

June 2017

12. The equation of a circle is $x^2 + y^2 - 12y + 20 = 0$. What are the coordinates of the center and the length of the radius of the circle?
(1) center (0, 6) and radius 4   (3) center (0, 6) and radius 16
(2) center (0, –6) and radius 4   (4) center (0, –6) and radius 16   12 ____

13. In the diagram of $\triangle RST$, $m\angle T = 90°$, $RS = 65$, and $ST = 60$.

What is the measure of $\angle S$, to the *nearest degree*?

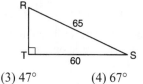

(1) 23°   (2) 43°   (3) 47°   (4) 67°   13 ____

14. Triangle $A'B'C'$ is the image of $\triangle ABC$ after a dilation followed by a translation. Which statement(s) would always be true with respect to this sequence of transformations?

I. $\triangle ABC \cong \triangle A'B'C'$

II. $\triangle ABC \sim \triangle A'B'C'$

III. $\overline{AB} \parallel \overline{A'B'}$

IV. $AA' = BB'$

(1) II and IV   (2) I and II   (3) II and III   (4) II, III, and IV   14 ____

15. Line segment RW has endpoints $R(-4, 5)$ and $W(6, 20)$. Point P is on $\overline{RW}$ such that $RP:PW$ is 2:3. What are the coordinates of point P?
(1) (2, 9)   (2) (0, 11)   (3) (2, 14)   (4) (10, 2)   15 ____

16. The pyramid shown has a square base, a height of 7, and a volume of 84.
What is the length of the side of the base?

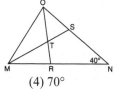

(1) 6   (3) 18
(2) 12   (4) 36   16 ____

17. In the diagram of triangle MNO, $\angle M$ and $\angle O$ are bisected by $\overline{MS}$ and $\overline{OR}$, respectively. Segments MS and OR intersect at T, and $m\angle N = 40°$.
If $m\angle TMR = 28°$, the measure of angle OTS is

(1) 40°   (2) 50°   (3) 60°   (4) 70°   17 ____

18. In the diagram, right triangle ABC has legs whose lengths are 4 and 6.

What is the volume of the three-dimensional object formed by continuously rotating the right triangle around $\overline{AB}$?

(1) 32π   (2) 48π   (3) 96π   (4) 144π   18 ____

19. What is an equation of a line that is perpendicular to the line whose equation is $2y = 3x - 10$ and passes through $(-6, 1)$?

(1) $y = -\frac{2}{3}x - 5$ (2) $y = -\frac{2}{3}x - 3$ (3) $y = \frac{2}{3}x + 1$ (4) $y = \frac{2}{3}x + 10$ 19 ____

20. In quadrilateral *BLUE* shown, $\overline{BE} \cong \overline{UL}$.

Which information would be sufficient to prove quadrilateral *BLUE* is a parallelogram?

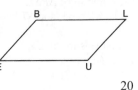

(1) $\overline{BL} \parallel \overline{EU}$ (3) $\overline{BE} \cong \overline{BL}$

(2) $\overline{LU} \parallel \overline{BE}$ (4) $\overline{LU} \cong \overline{EU}$ 20 ____

21. A ladder 20 feet long leans against a building, forming an angle of 71° with the level ground. To the *nearest foot*, how high up the wall of the building does the ladder touch the building?

(1) 15   (2) 16   (3) 18   (4) 19   21 ____

22. Which statement is sufficient evidence that $\triangle DEF$ is congruent to $\triangle ABC$?

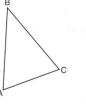
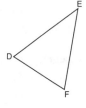

(1) $AB = DE$ and $BC = EF$

(2) $\angle D \cong \angle A$, $\angle B \cong \angle E$, $\angle C \cong \angle F$

(3) There is a sequence of rigid motions that maps $\overline{AB}$ onto $\overline{DE}$, $\overline{BC}$ onto $\overline{EF}$, and $\overline{AC}$ onto $\overline{DF}$.

(4) There is a sequence of rigid motions that maps point A onto point D, $\overline{AB}$ onto $\overline{DE}$, and $\angle B$ onto $\angle E$.   22 ____

23. A fabricator is hired to make a 27-foot-long solid metal railing for the stairs at the local library. The railing is modeled by the diagram. The railing is 2.5 inches high and 2.5 inches wide and is comprised of a rectangular prism and a half-cylinder.

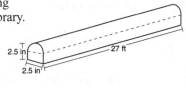

How much metal, to the *nearest cubic inch*, will the railing contain?

(1) 151   (2) 795   (3) 1808   (4) 2025   23 ____

24. In the diagram, $\triangle ABC \sim \triangle DEC$. If $AC = 12$, $DC = 7$, $DE = 5$, and the perimeter of $\triangle ABC$ is 30, what is the perimeter of $\triangle DEC$?

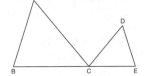

(1) 12.5   (3) 14.8

(2) 14.0   (4) 17.5   24 ____

Answer all 7 questions in this part. Each correct answer will receive 2 credits. Clearly indicate the necessary steps, including appropriate formula substitutions, diagrams, graphs, charts, etc. Utilize the information provided for each question to determine your answer. Note that diagrams are not necessarily drawn to scale. For all questions in this part, a correct numerical answer with no work shown will receive only 1 credit. All answers should be written in pen, except for graphs and drawings, which should he done in pencil. [14]

25. Given: Trapezoid *JKLM* with $\overline{JK} \parallel \overline{ML}$

Using a compass and straightedge, construct the altitude from vertex *J* to $\overline{ML}$. [Leave all construction marks.]

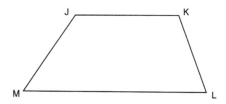

26. Determine and state, in terms of π, the area of a sector that intercepts a 40° arc of a circle with a radius of 4.5.

27. The diagram below shows two figures. Figure *A* is a right triangular prism and figure *B* is an oblique triangular prism. The base of figure *A* has a height of 5 and a length of 8 and the height of prism *A* is 14. The base of figure *B* has a height of 8 and a length of 5 and the height of prism *B* is 14.

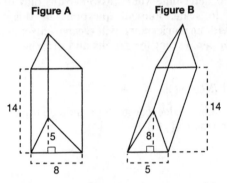

Use Cavalieri's Principle to explain why the volumes of these two triangular prisms are equal.

28. When volleyballs are purchased, they are not fully inflated. A partially inflated volleyball can be modeled by a sphere whose volume is approximately 180 in³. After being fully inflated, its volume is approximately 294 in³. To the *nearest tenth of an inch*, how much does the radius increase when the volleyball is fully inflated?

29. In right triangle *ABC* shown, altitude $\overline{CD}$ is drawn to hypotenuse $\overline{AB}$. Explain why $\triangle ABC \sim \triangle ACD$.

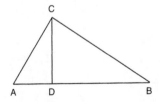

30. Triangle *ABC* and triangle *DEF* are drawn below.

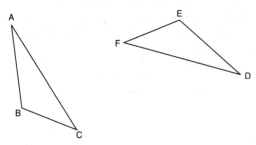

If $\overline{AB} \cong \overline{DE}$, $\overline{AC} \cong \overline{DF}$, and $\angle A \cong \angle D$, write a sequence of transformations that maps triangle *ABC* onto triangle *DEF*.

31. Line *n* is represented by the equation $3x + 4y = 20$. Determine and state the equation of line *p*, the image of line *n*, after a dilation of scale factor $\frac{1}{3}$ centered at the point (4, 2). [The use of the set of axes below is optional.]

Explain your answer.

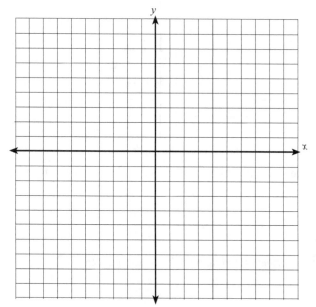

GEOMETRY
June 2017
Part III

Answer all **3** questions in this part. Each correct answer will receive **4 credits. Clearly indicate the necessary steps, including appropriate formula substitutions, diagrams, graphs, charts, etc. Utilize the information provided for each question to determine your answer. Note that diagrams are not necessarily drawn to scale. For all questions in this part, a correct numerical answer with no work shown will receive only 1 credit. All answers should be written in pen, except for graphs and drawings, which should be done in pencil.** [12]

32. Triangle *ABC* has vertices at *A*(–5, 2), *B*(–4, 7), and *C*(–2, 7), and triangle *DEF* has vertices at *D*(3, 2), *E*(2, 7), and *F*(0, 7). Graph and label △*ABC* and △*DEF* on the set of axes below.

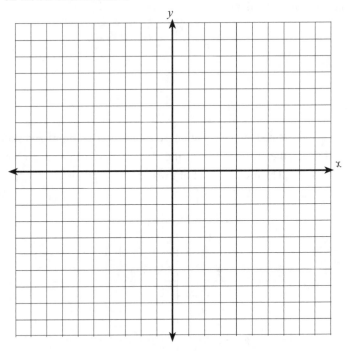

Determine and state the single transformation where △*DEF* is the image of △*ABC*.

Use your transformation to explain why △*ABC* ≅ △*DEF*.

33. Given: $\overline{RS}$ and $\overline{TV}$ bisect each other at point X

   $\overline{TR}$ and $\overline{SV}$ are drawn

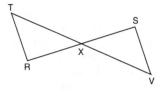

   Prove: $\overline{TR} \parallel \overline{SV}$

34. A gas station has a cylindrical fueling tank that holds the gasoline for its pumps, as modeled below. The tank holds a maximum of 20,000 gallons of gasoline and has a length of 34.5 feet.

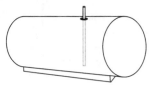

A metal pole is used to measure how much gas is in the tank. To the *nearest tenth of a foot*, how long does the pole need to be in order to reach the bottom of the tank and still extend one foot outside the tank? Justify your answer. [1 ft³ = 7.48 gallons]

Answer the 2 questions in this part. Each correct answer will receive 6 credits. Clearly indicate the necessary steps, including appropriate formula substitutions, diagrams, graphs, charts, etc. Utilize the information provided for each question to determine your answer. Note that diagrams are not necessarily drawn to scale. For all questions in this part, a correct numerical answer with no work shown will receive only 1 credit. All answers should be written in pen, except for graphs and drawings, which should be done in pencil. [12]

35. Quadrilateral *PQRS* has vertices *P*(–2, 3), *Q*(3, 8), *R*(4, 1), and *S*(–1, –4). Prove that *PQRS* is a rhombus. [The use of the set of axes is optional.]

Prove that *PQRS* is *not* a square. [The use of the set of axes is optional.]

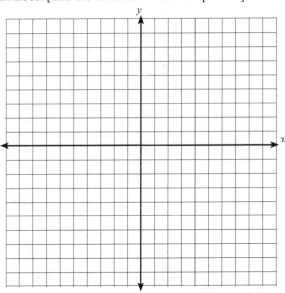

36. Freda, who is training to use a radar system, detects an airplane flying at a constant speed and heading in a straight line to pass directly over her location. She sees the airplane at an angle of elevation of 15° and notes that it is maintaining a constant altitude of 6250 feet. One minute later, she sees the airplane at an angle of elevation of 52°. How far has the airplane traveled, to the *nearest foot?*

Determine and state the speed of the airplane, to the *nearest mile per hour.*

Answer all 24 questions in this part. Each correct answer will receive 2 credits. Utilize the information provided for each question to determine your answer. Note that diagrams are not necessarily drawn to scale. For each statement or question, choose the word or expression that, of those given, best completes the statement or answers the question. [48]

1. A two-dimensional cross section is taken of a three-dimensional object. If this cross section is a triangle, what can *not* be the three-dimensional object?
(1) cone   (2) cylinder   (3) pyramid   (4) rectangular prism   1 _____

2. The image of $\triangle DEF$ is $\triangle D'E'F'$. Under which transformation will the triangles *not* be congruent?
(1) a reflection through the origin
(2) a reflection over the line $y = x$
(3) a dilation with a scale factor of 1 centered at $(2, 3)$
(4) a dilation with a scale factor of $\frac{3}{2}$ centered at the origin   2 _____

3. The vertices of square *RSTV* have coordinates $R(-1, 5)$, $S(-3, 1)$, $T(-7, 3)$, and $V(-5, 7)$. What is the perimeter of *RSTV*?
(1) $\sqrt{20}$ (2) $\sqrt{40}$ (3) $4\sqrt{20}$ (4) $4\sqrt{40}$ 3 _____

4. In the diagram of circle *O*, chord $\overline{CD}$ is parallel to diameter $\overline{AOB}$ and m$\overset{\frown}{CD}$= 130.
What is m$\overset{\frown}{AC}$?
(1) 25
(2) 50
(3) 65
(4) 115   4 _____

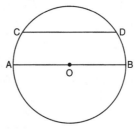

5. In the diagram below, $\overline{AD}$ intersects $\overline{BE}$ at C, and $\overline{AB} \parallel \overline{DE}$.

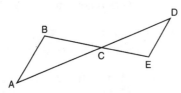

If $CD = 6.6$ cm, $DE = 3.4$ cm, $CE = 4.2$ cm, and $BC = 5.25$ cm, what is the length of $\overline{AC}$, to the *nearest hundredth of a centimeter*?
(1) 2.70   (2) 3.34   (3) 5.28   (4) 8.25   5 _____

August 2017

6. As shown in the graph, the quadrilateral is a rectangle.

Which transformation would *not* map the rectangle onto itself?
(1) a reflection over the *x*-axis
(2) a reflection over the line *x* = 4
(3) a rotation of 180° about the origin
(4) a rotation of 180° about the point (4, 0)

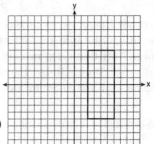

6 ____

7. In the diagram below, triangle *ACD* has points *B* and *E* on sides $\overline{AC}$ and $\overline{AD}$, respectively, such that $\overline{BE} \parallel \overline{CD}$, *AB* = 1, *BC* = 3.5, and *AD* = 18.

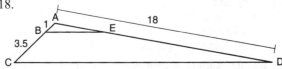

What is the length of $\overline{AE}$, to the *nearest tenth*?
(1) 14.0   (2) 5.1   (3) 3.3   (4) 4.0   7 ____

8. In the diagram of parallelogram *ROCK*, m∠*C* is 70° and m∠*ROS* is 65°.
What is m∠*KSO*?
(1) 45°   (3) 115°
(2) 110°   (4) 135°

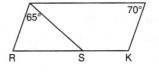

8 ____

9. In the diagram, ∠*GRS* ≅ ∠*ART*, *GR* = 36, *SR* = 45, *AR* = 15, and *RT* = 18.

Which triangle similarity statement is correct?

(1) △*GRS* ~ △*ART* by AA.   (3) △*GRS* ~ △*ART* by SSS.
(2) △*GRS* ~ △*ART* by SAS.   (4) △*GRS* is not similar to △*ART*.   9 ____

10. The line represented by the equation $4y = 3x + 7$ is transformed by a dilation centered at the origin. Which linear equation could represent its image?
(1) $3x - 4y = 9$ (3) $4x - 3y = 9$
(2) $3x + 4y = 9$ (4) $4x + 3y = 9$ 10 ____

11. Given △*ABC* with m∠*B* = 62° and side $\overline{AC}$ extended to *D*, as shown.

Which value of *x* makes $\overline{AB} \cong \overline{CB}$?
(1) 59°   (2) 62°   (3) 118°   (4) 121°

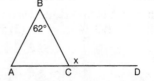

11 ____

12. In the diagram shown, $\overline{PA}$ is tangent to circle T at A, and secant $\overline{PBC}$ is drawn where point B is on circle T. If $PB = 3$ and $BC = 15$, what is the length of $\overline{PA}$?

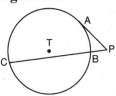

(1) $3\sqrt{5}$ (2) $3\sqrt{6}$ (3) 3   (4) 9   12 ____

13. A rectangle whose length and width are 10 and 6, respectively, is shown. The rectangle is continuously rotated around a straight line to form an object whose volume is 150π. Which line could the rectangle be rotated around?

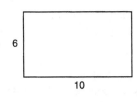

(1) a long side   (3) the vertical line of symmetry
(2) a short side   (4) the horizontal line of symmetry   13 ____

14. If $ABCD$ is a parallelogram, which statement would prove that $ABCD$ is a rhombus?

(1) $\angle ABC \cong \angle CDA$ (3) $\overline{AC} \perp \overline{BD}$
(2) $\overline{AC} \cong \overline{BD}$ (4) $\overline{AB} \perp \overline{CD}$ 14 ____

15. To build a handicapped-access ramp, the building code states that for every 1 inch of vertical rise in height, the ramp must extend out 12 inches horizontally, as shown in the diagram below.

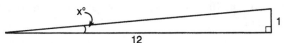

What is the angle of inclination, x, of this ramp, to the *nearest hundredth of a degree*?

(1) 4.76   (2) 4.78   (3) 85.22   (4) 85.24   15 ____

16. In the diagram of $\triangle ABC$, D, E, and F are the midpoints of $\overline{AB}$, $\overline{BC}$, and $\overline{CA}$, respectively. What is the ratio of the area of $\triangle CFE$ to the area of $\triangle CAB$?

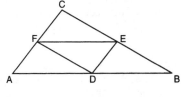

(1) 1:1   (3) 1:3
(2) 1:2   (4) 1:4   16 ____

17. The coordinates of the endpoints of $\overline{AB}$ are $A(-8, -2)$ and $B(16, 6)$. Point P is on $\overline{AB}$. What are the coordinates of point P, such that $AP:PB$ is 3:5?

(1) (1, 1)   (2) (7, 3)   (3) (9.6, 3.6)   (4) (6.4, 2.8)   17 ____

18. Kirstie is testing values that would make triangle *KLM* a right triangle when *LN* is an altitude, and *KM* = 16, as shown.

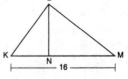

Which lengths would make triangle *KLM* a right triangle?
(1) *LM* = 13 and *KN* = 6   (3) *KL* = 11 and *KN* = 7
(2) *LM* = 12 and *NM* = 9   (4) *LN* = 8 and *NM* = 10   18 ____

19. In right triangle *ABC*, m∠*A* = 32°, m∠*B* = 90°, and *AC* = 6.2 cm. What is the length of $\overline{BC}$, to the *nearest tenth of a centimeter*?
(1) 3.3   (2) 3.9   (3) 5.3   (4) 11.7   19 ____

20. The 2010 U.S. Census populations and population densities are shown in the table.

State	Population Density $\left(\frac{people}{mi^2}\right)$	Population in 2010
Florida	350.6	18,801,310
Illinois	231.1	12,830,632
New York	411.2	19,378,102
Pennsylvania	283.9	12,702,379

Based on the table above, which list has the states' areas, in square miles, in order from largest to smallest?
(1) Illinois, Florida, New York, Pennsylvania
(2) New York, Florida, Illinois, Pennsylvania
(3) New York, Florida, Pennsylvania, Illinois
(4) Pennsylvania, New York, Florida, Illinois   20 ____

21. In a right triangle, sin (40 − *x*)° = cos (3*x*)°. What is the value of *x*?
(1) 10   (2) 15   (3) 20   (4) 25   21 ____

22. A regular decagon is rotated *n* degrees about its center, carrying the decagon onto itself. The value of *n* could be
(1) 10°   (2) 150°   (3) 225°   (4) 252°   22 ____

23. In a circle with a diameter of 32, the area of a sector is $\frac{512\pi}{3}$. The measure of the angle of the sector, in radians, is
(1) $\frac{\pi}{3}$ (2) $\frac{4\pi}{3}$ (3) $\frac{16\pi}{3}$ (4) $\frac{64\pi}{3}$ 23 ____

24. What is an equation of the perpendicular bisector of the line segment shown in the diagram?
(1) *y* + 2*x* = 0
(2) *y* − 2*x* = 0
(3) 2*y* + *x* = 0
(4) 2*y* − *x* = 0

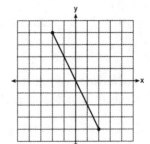

24 ____

Answer all 7 questions in this part. Each correct answer will receive 2 credits. Clearly indicate the necessary steps, including appropriate formula substitutions, diagrams, graphs, charts, etc. Utilize the information provided for each question to determine your answer. Note that diagrams are not necessarily drawn to scale. For all questions in this part, a correct numerical answer with no work shown will receive only 1 credit. All answers should be written in pen, except for graphs and drawings, which should be done in pencil. [14]

25. Sue believes that the two cylinders shown in the diagram have equal volumes.

Is Sue correct? Explain why.

26. In the diagram of rhombus $PQRS$, the diagonals $\overline{PR}$ and $\overline{QS}$ intersect at point T, $PR = 16$, and $QS = 30$. Determine and state the perimeter of $PQRS$.

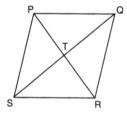

27. Quadrilateral $MATH$ and its image $M''A''T''H''$ are graphed on the set of axes.

Describe a sequence of transformations that maps quadrilateral $MATH$ onto quadrilateral $M''A''T''H''$.

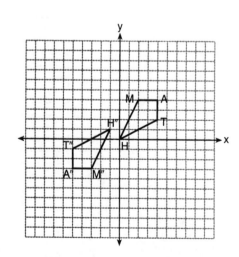

GEOMETRY
August 2017

28. Using a compass and straightedge, construct a regular hexagon inscribed in circle O. [Leave all construction marks.]

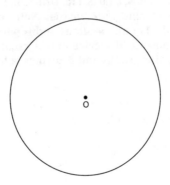

29. The coordinates of the endpoints of $\overline{AB}$ are $A(2, 3)$ and $B(5, -1)$. Determine the length of $\overline{A'B'}$, the image of $\overline{AB}$, after a dilation of $\frac{1}{2}$ centered at the origin. [The use of the set of axes below is optional.]

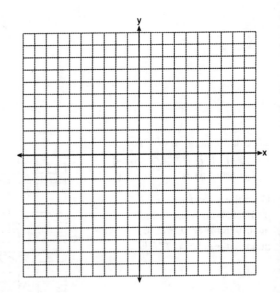

30. In the diagram below of $\triangle ABC$ and $\triangle XYZ$, a sequence of rigid motions maps $\angle A$ onto $\angle X$, $\angle C$ onto $\angle Z$, and $\overline{AC}$ onto $\overline{XZ}$.

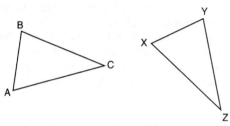

Determine and state whether $\overline{BC} \cong \overline{YZ}$. Explain why.

31. Determine and state the coordinates of the center and the length of the radius of a circle whose equation is $x^2 + y^2 - 6x = 56 - 8y$.

GEOMETRY
August 2017
Part III

Answer all 3 questions in this part. Each correct answer will receive 4 credits. Clearly indicate the necessary steps, including appropriate formula substitutions, diagrams, graphs, charts, etc. Utilize the information provided for each question to determine your answer. Note that diagrams are not necessarily drawn to scale. For all questions in this part, a correct numerical answer with no work shown will receive only 1 credit. All answers should be written in pen, except for graphs and drawings, which should be done in pencil. [12]

32. Triangle PQR has vertices $P(-3, -1)$, $Q(-1, 7)$, and $R(3, 3)$, and points A and B are midpoints of $\overline{PQ}$ and $\overline{RQ}$, respectively. Use coordinate geometry to prove that $\overline{AB}$ is parallel to $\overline{PR}$ and is half the length of $\overline{PR}$.

[The use of the set of axes below is optional.]

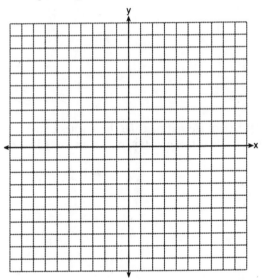

33. In the diagram of circle O, tangent $\overline{EC}$ is drawn to diameter $\overline{AC}$. Chord $\overline{BC}$ is parallel to secant $\overline{ADE}$, and chord $\overline{AB}$ is drawn.

Prove: $\dfrac{BC}{CA} = \dfrac{AB}{EC}$

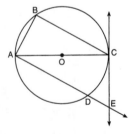

34. Keira has a square poster that she is framing and placing on her wall. The poster has a diagonal 58 cm long and fits exactly inside the frame. The width of the frame around the picture is 4 cm.

Determine and state the total area of the poster and frame to the *nearest tenth of a square centimeter*.

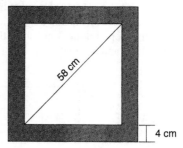

58 cm

4 cm

Part IV
Answer the 2 questions in this part. Each correct answer will receive 6 credits. Clearly indicate the necessary steps, including appropriate formula substitutions, diagrams, graphs, charts, etc. Utilize the information provided for each question to determine your answer. Note that diagrams are not necessarily drawn to scale. For all questions in this part, a correct numerical answer with no work shown will receive only 1 credit. All answers should be written in pen, except for graphs and drawings, which should be done in pencil. [12]

35. Isosceles trapezoid *ABCD* has bases $\overline{DC}$ and $\overline{AB}$ with nonparallel legs $\overline{AD}$ and $\overline{BC}$. Segments *AE*, *BE*, *CE*, and *DE* are drawn in trapezoid *ABCD* such that $\angle CDE \cong \angle DCE$, $\overline{AE} \perp \overline{DE}$, and $\overline{BE} \perp \overline{CE}$.

Prove $\triangle ADE \cong \triangle BCE$ and prove $\triangle AEB$ is an isosceles triangle.

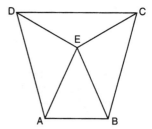

36. A rectangular in-ground pool is modeled by the prism below. The inside of the pool is 16 feet wide and 35 feet long. The pool has a shallow end and a deep end, with a sloped floor connecting the two ends. Without water, the shallow end is 9 feet long and 4.5 feet deep, and the deep end of the pool is 12.5 feet long.

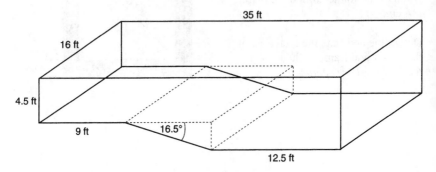

If the sloped floor has an angle of depression of 16.5 degrees, what is the depth of the pool at the deep end, to the *nearest tenth of a foot*?

Find the volume of the inside of the pool to the *nearest cubic foot*.

A garden hose is used to fill the pool. Water comes out of the hose at a rate of 10.5 gallons per minute. How much time, to the *nearest hour*, will it take to fill the pool 6 inches from the top? [1 ft³ = 7.48 gallons]

January 2018
Part I

Answer all 24 questions in this part. Each correct answer will receive 2 credits. No partial credit will be allowed. Utilize the information provided for each question to determine your answer. Note that diagrams are not necessarily drawn to scale. For each statement or question, choose the word or expression that, of those given, best completes the statement or answers the question. Record your answers in the space provided [48]

1. In the diagram, a sequence of rigid motions maps *ABCD* onto *JKLM*. If m∠*A* = 82°, m∠*B* = 104°, and m∠*L* = 121°, the measure of ∠*M* is
(1) 53°
(2) 82°
(3) 104°
(4) 121°

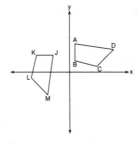

1 _____

2. Parallelogram *HAND* is drawn with diagonals $\overline{HN}$ and $\overline{AD}$ intersecting at *S*. Which statement is always true?

(1) $HN = \frac{1}{2}AD$

(2) $AS = \frac{1}{2}AD$

(3) ∠*AHS* ≅ ∠*ANS*

(4) ∠*HDS* ≅ ∠*NDS*

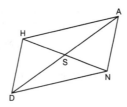

2 _____

3. The graph shows two congruent triangles, *ABC* and *A'B'C'*. Which rigid motion would map △*ABC* onto △*A'B'C'*?
(1) a rotation of 90 degrees counterclockwise about the origin
(2) a translation of three units to the left and three units up
(3) a rotation of 180 degrees about the origin
(4) a reflection over the line *y* = *x*

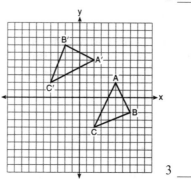

3 _____

4. A man was parasailing above a lake at an angle of elevation of 32° from a boat, as modeled in the diagram. If 129.5 meters of cable connected the boat to the parasail, approximately how many meters above the lake was the man?
(1) 68.6
(2) 80.9
(3) 109.8
(4) 244.4

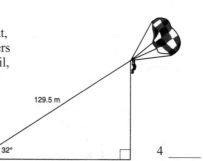

129.5 m

32°

4 _____

5. A right hexagonal prism is shown. A two-dimensional
cross section that is perpendicular to the base is taken
from the prism.

Which figure describes the two-dimensional cross section?
(1) triangle   (2) rectangle   (3) pentagon   (4) hexagon   5 _____

6. In the diagram, $\overline{AC}$ has endpoints
with coordinates $A(-5, 2)$ and $C(4, -10)$.

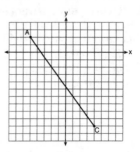

If B is a point on $\overline{AC}$ and $AB:BC = 1:2$,
what are the coordinates of B?

(1) $(-2, -2)$ (3) $(0, -\frac{14}{3})$

(2) $(-\frac{1}{2}, -4)$ (4) $(1, -6)$ 6 _____

7. An ice cream waffle cone can be modeled by a right circular cone with a
base diameter of 6.6 centimeters and a volume of 54.45π cubic centimeters.
What is the number of centimeters in the height of the waffle cone?

(1) $3\frac{3}{4}$ (2) 5   (3) 15   (4) $24\frac{3}{4}$ 7 _____

8. The vertices of $\triangle PQR$ have coordinates $P(2, 3)$, $Q(3, 8)$, and $R(7, 3)$.
Under which transformation of $\triangle PQR$ are distance and angle measure
preserved?

(1) $(x, y) \rightarrow (2x, 3y)$ (3) $(x, y) \rightarrow (2x, y + 3)$
(2) $(x, y) \rightarrow (x + 2, 3y)$ (4) $(x, y) \rightarrow (x + 2, y + 3)$ 8 _____

9. In $\triangle ABC$ shown below, side $\overline{AC}$ is extended to point D with
$m\angle DAB = (180 - 3x)°$, $m\angle B = (6x - 40)°$, and $m\angle C = (x + 20)°$.

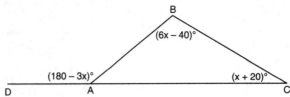

What is $m\angle BAC$?
(1) $20°$ (2) $40°$ (3) $60°$ (4) $80°$ 9 _____

10. Circle O is centered at the origin.
In the diagram, a quarter of circle O
is graphed.

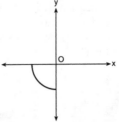

Which three-dimensional figure is
generated when the quarter circle is
continuously rotated about the y-axis?
(1) cone   (3) cylinder
(2) sphere   (4) hemisphere   10 _____

11. Rectangle $A'B'C'D'$ is the image of rectangle $ABCD$ after a dilation centered at point A by a scale factor of $\frac{2}{3}$. Which statement is correct?

(1) Rectangle $A'B'C'D'$ has a perimeter that is $\frac{2}{3}$ the perimeter of rectangle $ABCD$.

(2) Rectangle $A'B'C'D'$ has a perimeter that is $\frac{3}{2}$ the perimeter of rectangle $ABCD$.

(3) Rectangle $A'B'C'D'$ has an area that is $\frac{2}{3}$ the area of rectangle $ABCD$.

(4) Rectangle $A'B'C'D'$ has an area that is $\frac{3}{2}$ the area of rectangle $ABCD$.

11 _____

12. The equation of a circle is $x^2 + y^2 - 6x + 2y = 6$. What are the coordinates of the center and the length of the radius of the circle?
(1) center $(-3, 1)$ and radius 4   (3) center $(-3, 1)$ and radius 16
(2) center $(3, -1)$ and radius 4   (4) center $(3, -1)$ and radius 16   12 _____

13. In the diagram of $\triangle ABC$, $\overline{DE}$ is parallel to $\overline{AB}$, $CD = 15$, $AD = 9$, and $AB = 40$. The length of $\overline{DE}$ is
(1) 15   (3) 25
(2) 24   (4) 30

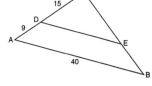

13 _____

14. The line whose equation is $3x - 5y = 4$ is dilated by a scale factor of $\frac{5}{3}$ centered at the origin. Which statement is correct?
(1) The image of the line has the same slope as the pre-image but a different y-intercept.
(2) The image of the line has the same y-intercept as the pre-image but a different slope.
(3) The image of the line has the same slope and the same y-intercept as the pre-image.
(4) The image of the line has a different slope and a different y-intercept from the pre-image.

14 _____

15. Which transformation would *not* carry a square onto itself?
(1) a reflection over one of its diagonals
(2) a 90° rotation clockwise about its center
(3) a 180° rotation about one of its vertices
(4) a reflection over the perpendicular bisector of one side

15 _____

16. In circle M, diameter $\overline{AC}$, chords $\overline{AB}$ and $\overline{BC}$, and radius $\overline{MB}$ are drawn.
Which statement is *not* true?
(1) $\triangle ABC$ is a right triangle.
(2) $\triangle ABM$ is isosceles.
(3) $m\widehat{BC} = m\angle BMC$
(4) $m\widehat{AB} = \frac{1}{2}m\angle ACB$

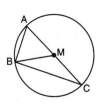

16 _____

17. In the diagram, $\overline{XS}$ and $\overline{YR}$ intersect at Z. Segments XY and RS are drawn perpendicular to $\overline{YR}$ to form triangles XYZ and SRZ.

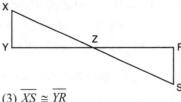

Which statement is always true?
(1) $(XY)(SR) = (XZ)(RZ)$
(3) $\overline{XS} \cong \overline{YR}$
(2) $\triangle XYZ \cong \triangle SRZ$
(4) $\dfrac{XY}{SR} = \dfrac{YZ}{RZ}$

17 _____

18. As shown in the diagram, $\overleftrightarrow{ABC} \parallel \overleftrightarrow{EFG}$ and $\overline{BF} \cong \overline{EF}$.

If m$\angle CBF = 42.5°$, then m$\angle EBF$ is
(1) $42.5°$
(2) $68.75°$
(3) $95°$
(4) $137.5°$

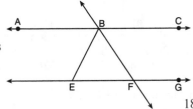

18 _____

19. A parallelogram must be a rhombus if its diagonals
(1) are congruent
(3) do not bisect its angles
(2) bisect each other
(4) are perpendicular to each other

19 _____

20. What is an equation of a line which passes through $(6, 9)$ and is perpendicular to the line whose equation is $4x - 6y = 15$?

(1) $y - 9 = -\dfrac{3}{2}(x - 6)$
(3) $y + 9 = -\dfrac{3}{2}(x + 6)$
(2) $y - 9 = \dfrac{2}{3}(x - 6)$
(4) $y + 9 = \dfrac{2}{3}(x + 6)$

20 _____

21. Quadrilateral $ABCD$ is inscribed in circle O, as shown. If m$\angle A = 80°$, m$\angle B = 75°$, m$\angle C = (y + 30)°$, and m$\angle D = (x - 10)°$, which statement is true?
(1) $x = 85$ and $y = 50$
(2) $x = 90$ and $y = 45$
(3) $x = 110$ and $y = 75$
(4) $x = 115$ and $y = 70$

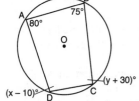

21 _____

22. A regular pyramid has a square base. The perimeter of the base is 36 inches and the height of the pyramid is 15 inches. What is the volume of the pyramid in cubic inches?
(1) 180
(2) 405
(3) 540
(4) 1215

22 _____

23. In the diagram of $\triangle ABC$, $\angle ABC$ is a right angle, $AC = 12$, $AD = 8$, and altitude $\overline{BD}$ is drawn. What is the length of $\overline{BC}$?

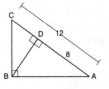

(1) $4\sqrt{2}$ (2) $4\sqrt{3}$ (3) $4\sqrt{5}$ (4) $4\sqrt{6}$

23 _____

24. In the diagram, two concentric circles with center O, and radii $\overline{OC}$, $\overline{OD}$, $\overline{OCE}$, and $\overline{ODF}$ are drawn. If $OC = 4$ and $OE = 6$, which relationship between the length of arc EF and the length of arc CD is always true?

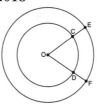

(1) The length of arc EF is 2 units longer than the length of arc CD.
(2) The length of arc EF is 4 units longer than the length of arc CD.
(3) The length of arc EF is 1.5 times the length of arc CD.
(4) The length of arc EF is 2.0 times the length of arc CD.

24 ____

Part II

Answer all 7 questions in this part. Each correct answer will receive 2 credits. Clearly indicate the necessary steps, including appropriate formula substitutions, diagrams, graphs, charts, etc. Utilize the information provided for each question to determine your answer. Note that diagrams are not necessarily drawn to scale. For all questions in this part, a correct numerical answer with no work shown will receive only 1 credit. All answers should be written in pen, except for graphs and drawings, which should be done in pencil. [14]

25. Given: Parallelogram $ABCD$ with diagonal $\overline{AC}$ drawn

Prove: $\triangle ABC \cong \triangle CDA$

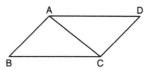

26. The diagram below shows circle O with diameter $\overline{AB}$. Using a compass and straightedge, construct a square that is inscribed in circle O.
[Leave all construction marks.]

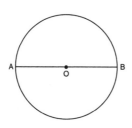

27. Given: Right triangle ABC with right angle at C
If sin A increases, does cos B increase or decrease? Explain why.

28. In the diagram, the circle has a radius of 25 inches. The area of the *unshaded* sector is 500π in^2.

Determine and state the degree measure of angle Q, the central angle of the shaded sector.

29. A machinist creates a solid steel part for a wind turbine engine. The part has a volume of 1015 cubic centimeters. Steel can be purchased for $0.29 per kilogram, and has a density of 7.95 g/cm^3.

If the machinist makes 500 of these parts, what is the cost of the steel, to the *nearest dollar*?

30. In the graph, △*ABC* has coordinates *A*(–9, 2), *B*(–6, –6), and *C*(–3, –2), and △*RST* has coordinates *R*(–2, 9), *S*(5, 6), and *T*(2, 3).

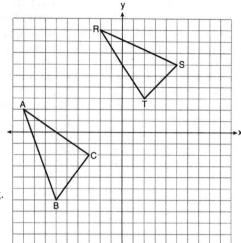

Is △*ABC* congruent to △*RST*? Use the properties of rigid motions to explain your reasoning.

31. Bob places an 18-foot ladder 6 feet from the base of his house and leans it up against the side of his house. Find, to the *nearest degree*, the measure of the angle the bottom of the ladder makes with the ground.

GEOMETRY
January 2018
Part III

Answer all 3 questions in this part. Each correct answer will receive 4 credits. Clearly indicate the necessary steps, including appropriate formula substitutions, diagrams, graphs, charts, etc. Utilize the information provided for each question to determine your answer. Note that diagrams are not necessarily drawn to scale. For all questions in this part, a correct numerical answer with no work shown will receive only 1 credit. All answers should be written in pen, except for graphs and drawings, which should be done in pencil. [12]

32. Triangle *ABC* and triangle *ADE* are graphed on the set of axes.

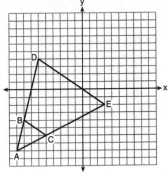

Describe a transformation that maps triangle *ABC* onto triangle *ADE*.

Explain why this transformation makes triangle *ADE* similar to triangle *ABC*.

33. A storage tank is in the shape of a cylinder with a hemisphere on the top. The highest point on the inside of the storage tank is 13 meters above the floor of the storage tank, and the diameter inside the cylinder is 8 meters. Determine and state, to the *nearest cubic meter*, the total volume inside the storage tank.

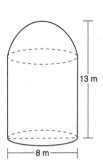

34. As shown in the diagram, an island (*I*) is due north of a marina (*M*). A boat house (*H*) is 4.5 miles due west of the marina. From the boat house, the island is located at an angle of 54° from the marina.

Determine and state, to the *nearest tenth of a mile*, the distance from the boat house (*H*) to the island (*I*).

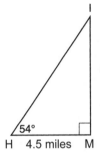

Determine and state, to the *nearest tenth of a mile*, the distance from the island (*I*) to the marina (*M*).

GEOMETRY
January 2018
Part IV

Answer the question in this part. A correct answer will receive 6 credits. Clearly indicate the necessary steps, including appropriate formula substitutions, diagrams, graphs, charts, etc. Utilize the information provided for each question to determine your answer. Note that diagrams are not necessarily drawn to scale. A correct numerical answer with no work shown will receive only 1 credit. All answers should be written in pen, except for graphs and drawings, which should be done in pencil. [6]

35. In the coordinate plane, the vertices of triangle PAT are $P(-1, -6)$, $A(-4, 5)$, and $T(5, -2)$. Prove that $\triangle PAT$ is an isosceles triangle. [The use of the set of axes is optional.]

State the coordinates of R so that quadrilateral $PART$ is a parallelogram.

Prove that quadrilateral $PART$ is a parallelogram.

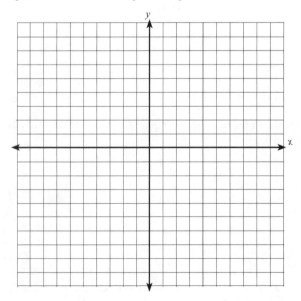

June 2018
Part I

Answer all **24** questions in this part. Each correct answer will receive **2** credits. No partial credit will be allowed. Utilize the information provided for each question to determine your answer. Note that diagrams are not necessarily drawn to scale. For each statement or question, choose the word or expression that, of those given, best completes the statement or answers the question. Record your answers in the space provided [48]

1. After a counterclockwise rotation about point X, scalene triangle ABC maps onto $\triangle RST$, as shown in the diagram. Which statement must be true?

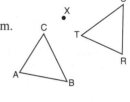

(1) $\angle A \cong \angle R$ (3) $\overline{CB} \cong \overline{TR}$
(2) $\angle A \cong \angle S$ (4) $\overline{CA} \cong \overline{TS}$

1 _____

2. In the diagram, $\overrightarrow{AB} \parallel \overrightarrow{DEF}$, $\overline{AE}$ and $\overline{BD}$ intersect at C, $m\angle B = 43°$, and $m\angle CEF = 152°$.

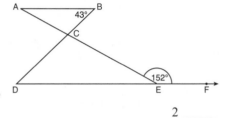

Which statement is true?
(1) $m\angle D = 28°$ (3) $m\angle ACD = 71°$
(2) $m\angle A = 43°$ (4) $m\angle BCE = 109°$

2 _____

3. In the diagram, line m is parallel to line n. Figure 2 is the image of Figure 1 after a reflection over line m. Figure 3 is the image of Figure 2 after a reflection over line n.

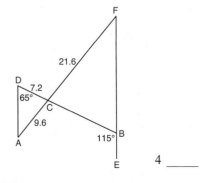

Which single transformation would carry Figure 1 onto Figure 3?
(1) a dilation   (3) a reflection
(2) a rotation   (4) a translation

3 _____

4. In the diagram, $\overline{AF}$ and $\overline{DB}$ intersect at C, and $\overline{AD}$ and $\overline{FBE}$ are drawn such that $m\angle D = 65°$, $m\angle CBE = 115°$, $DC = 7.2$, $AC = 9.6$, and $FC = 21.6$.

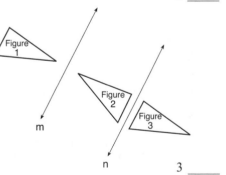

What is the length of $\overline{CB}$?
(1) 3.2
(2) 4.8
(3) 16.2
(4) 19.2

4 _____

5. Given square RSTV, where RS = 9 cm. If square RSTV is dilated by a scale factor of 3 about a given center, what is the perimeter, in centimeters, of the image of RSTV after the dilation?

(1) 12   (2) 27   (3) 36   (4) 108   5 _____

6. In right triangle ABC, hypotenuse $\overline{AB}$ has a length of 26 cm, and side $\overline{BC}$ has a length of 17.6 cm. What is the measure of angle B, to the *nearest degree*?

(1) 48°   (2) 47°   (3) 43°   (4) 34°   6 _____

7. The greenhouse pictured can be modeled as a rectangular prism with a half-cylinder on top. The rectangular prism is 20 feet wide, 12 feet high, and 45 feet long. The half-cylinder has a diameter of 20 feet. To the *nearest cubic foot*, what is the volume of the greenhouse?

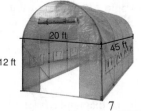

(1) 17,869   (3) 39,074
(2) 24,937   (4) 67,349   7 _____

8. In a right triangle, the acute angles have the relationship $\sin(2x+4) = \cos(46)$. What is the value of x?

(1) 20   (2) 21   (3) 24   (4) 25   8 _____

9. In the diagram $\overline{AB} \parallel \overline{DFC}$, $\overline{EDA} \parallel \overline{CBG}$, and $\overline{EFB}$ and $\overline{AG}$ are drawn.

Which statement is always true?

(1) $\triangle DEF \cong \triangle CBF$

(2) $\triangle BAG \cong \triangle BAE$

(3) $\triangle BAG \sim \triangle AEB$

(4) $\triangle DEF \sim \triangle AEB$

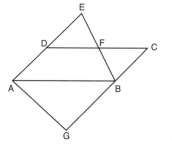

                                      9 _____

10. The base of a pyramid is a rectangle with a width of 4.6 cm and a length of 9 cm. What is the height, in centimeters, of the pyramid if its volume is 82.8 cm³?

(1) 6   (2) 2   (3) 9   (4) 18   10 _____

11. In the diagram of right triangle AED, $\overline{BC} \parallel \overline{DE}$. Which statement is always true?

(1) $\dfrac{AC}{BC} = \dfrac{DE}{AE}$ (3) $\dfrac{AC}{CE} = \dfrac{BC}{DE}$

(2) $\dfrac{AB}{AD} = \dfrac{BC}{DE}$ (4) $\dfrac{DE}{BC} = \dfrac{DB}{AB}$

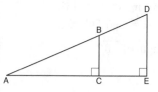

                                    11 _____

12. What is an equation of the line that passes through the point (6, 8) and is perpendicular to a line with equation $y = \frac{3}{2}x + 5$?

(1) $y - 8 = \frac{3}{2}(x - 6)$

(3) $y + 8 = \frac{3}{2}(x + 6)$

(2) $y - 8 = -\frac{2}{3}(x - 6)$

(4) $y + 8 = -\frac{2}{3}(x + 6)$

12 ____

13. The diagram shows parallelogram $ABCD$ with diagonals $\overline{AC}$ and $\overline{BD}$ intersecting at E. What additional information is sufficient to prove that parallelogram $ABCD$ is also a rhombus?

(1) $\overline{BD}$ bisects $\overline{AC}$.

(3) $\overline{AC}$ is congruent to $\overline{BD}$.

(2) $\overline{AB}$ is parallel to $\overline{CD}$.

(4) $\overline{AC}$ is perpendicular to $\overline{BD}$.

13 ____

14. Directed line segment DE has endpoints $D(-4, -2)$ and $E(1, 8)$. Point F divides $\overline{DE}$ such that $DF:FE$ is 2:3. What are the coordinates of F?

(1) $(-3, 0)$ (2) $(-2, 2)$ (3) $(-1, 4)$ (4) $(2, 4)$ 14 ____

15. Triangle DAN is graphed on the set of axes. The vertices of $\triangle DAN$ have coordinates $D(-6, -1)$, $A(6, 3)$, and $N(-3, 10)$.

What is the area of $\triangle DAN$?

(1) 60

(3) $20\sqrt{13}$

(2) 120

(4) $40\sqrt{13}$

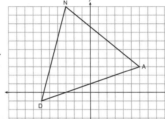

15 ____

16. Triangle ABC, with vertices at $A(0, 0)$, $B(3, 5)$, and $C(0, 5)$, is graphed on the set of axes shown.

Which figure is formed when $\triangle ABC$ is rotated continuously about $\overline{BC}$?

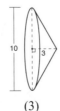

(1)   (2)   (3)   (4)   16 ____

17. In the diagram of circle O, chords $\overline{AB}$ and $\overline{CD}$ intersect at E. If $m\overarc{AC} = 72°$ and $m\angle AEC = 58°$, how many degrees are in $m\overarc{DB}$?

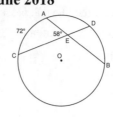

(1) 108°   (3) 44°
(2) 65°   (4) 14°

17 ____

18. In triangle SRK, medians $\overline{SC}$, $\overline{KE}$, and $\overline{RL}$ intersect at M. Which statement must always be true?

(1) $3(MC) = SC$ (3) $RM = 2MC$
(2) $MC = \frac{1}{3}(SM)$ (4) $SM = KM$

18 ____

19. The regular polygon is rotated about its center. Which angle of rotation will carry the figure onto itself?

(1) 60°   (3) 216°
(2) 108°   (4) 540°

19 ____

20. What is an equation of circle O shown in the graph?

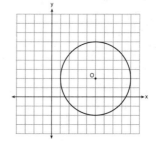

(1) $x^2 + 10x + y^2 + 4y = -13$

(2) $x^2 - 10x + y^2 - 4y = -13$

(3) $x^2 + 10x + y^2 + 4y = -25$

(4) $x^2 - 10x + y^2 - 4y = -25$

20 ____

21. In the diagram of $\triangle PQR$, $\overline{ST}$ is drawn parallel to $\overline{PR}$, $PS = 2$, $SQ = 5$, and $TR = 5$. What is the length of $\overline{QR}$?

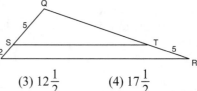

(1) 7   (2) 2   (3) $12\frac{1}{2}$ (4) $17\frac{1}{2}$

21 ____

22. The diagram shows circle O with radii $\overline{OA}$ and $\overline{OB}$. The measure of angle AOB is 120°, and the length of a radius is 6 inches. Which expression represents the length of arc AB, in inches?

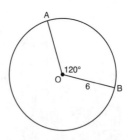

(1) $\frac{120}{360}(6\pi)$ (3) $\frac{1}{3}(36\pi)$

(2) $120(6)$ (4) $\frac{1}{3}(12\pi)$

22 ____

23. Line segment CD is the altitude drawn to hypotenuse $\overline{EF}$ in right triangle ECF. If $EC = 10$ and $EF = 24$, then, to the *nearest tenth*, ED is

(1) 4.2   (2) 5.4   (3) 15.5   (4) 21.8   23 ____

24. Line MN is dilated by a scale factor of 2 centered at the point $(0, 6)$. If $\overleftrightarrow{MN}$ is represented by $y = -3x + 6$, which equation can represent $\overleftrightarrow{M'N'}$, the image of $\overleftrightarrow{MN}$?

(1) $y = -3x + 12$ (2) $y = -3x + 6$ (3) $y = -6x + 12$ (4) $y = -6x + 6$ 24 ____

Part II

Answer all 7 questions in this part. Each correct answer will receive 2 credits. Clearly indicate the necessary steps, including appropriate formula substitutions, diagrams, graphs, charts, etc. Utilize the information provided for each question to determine your answer. Note that diagrams are not necessarily drawn to scale. For all questions in this part, a correct numerical answer with no work shown will receive only 1 credit. All answers should be written in pen, except for graphs and drawings, which should be done in pencil. [14]

25. Triangle $A'B'C'$ is the image of triangle ABC after a translation of 2 units to the right and 3 units up. Is triangle ABC congruent to triangle $A'B'C'$? Explain why.

26. Triangle ABC and point $D(1, 2)$ are graphed on the set of axes. Graph and label $\triangle A'B'C'$, the image of $\triangle ABC$, after a dilation of scale factor 2 centered at point D.

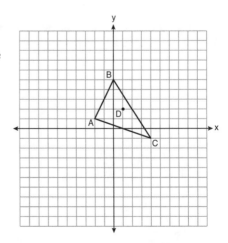

June 2018

27. Quadrilaterals *BIKE* and *GOLF* are graphed on the set of axes.

Describe a sequence of transformations that maps quadrilateral *BIKE* onto quadrilateral *GOLF*.

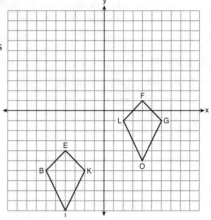

28. In the diagram, secants $\overline{RST}$ and $\overline{RQP}$, drawn from point R, intersect circle O at S, T, Q, and P.

If $RS = 6$, $ST = 4$, and $RP = 15$, what is the length of $\overline{RQ}$?

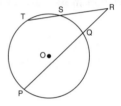

29. Using a compass and straightedge, construct the median to side $\overline{AC}$ in $\triangle ABC$ below. [Leave all construction marks.]

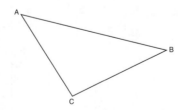

June 2018

30. Skye says that the two triangles below are congruent. Margaret says that the two triangles are similar.

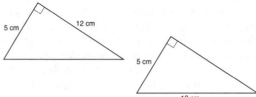

Are Skye and Margaret both correct? Explain why.

31. Randy's basketball is in the shape of a sphere with a maximum circumference of 29.5 inches. Determine and state the volume of the basketball, to the *nearest cubic inch*.

GEOMETRY
June 2018
Part III

Answer all 3 questions in this part. Each correct answer will receive 4 credits. Clearly indicate the necessary steps, including appropriate formula substitutions, diagrams, graphs, charts, etc. Utilize the information provided for each question to determine your answer. Note that diagrams are not necessarily drawn to scale. For all questions in this part, a correct numerical answer with no work shown will receive only 1 credit. All answers should be written in pen, except for graphs and drawings, which should be done in pencil. [12]

32. Triangle *ABC* has vertices with coordinates $A(-1, -1)$, $B(4, 0)$, and $C(0, 4)$. Prove that $\triangle ABC$ is an isosceles triangle but *not* an equilateral triangle. [The use of the set of axes below is optional.]

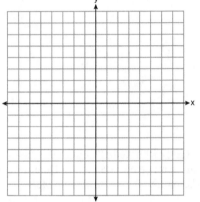

33. The map of a campground is shown below. Campsite *C*, first aid station *F*, and supply station *S* lie along a straight path. The path from the supply station to the tower, *T*, is perpendicular to the path from the supply station to the campsite. The length of path $\overline{FS}$ is 400 feet. The angle formed by path $\overline{TF}$ and path $\overline{FS}$ is 72°. The angle formed by path $\overline{TC}$ and path $\overline{CS}$ is 55°.

Determine and state, to the *nearest foot*, the distance from the campsite to the tower.

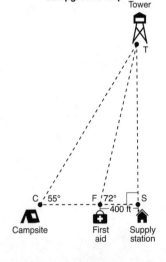

Campground Map

34. Shae has recently begun kickboxing and purchased training equipment as modeled in the diagram. The total weight of the bag, pole, and unfilled base is 270 pounds. The cylindrical base is 18 inches tall with a diameter of 20 inches. The dry sand used to fill the base weighs 95.46 lbs per cubic foot.

To the *nearest pound*, determine and state the total weight of the training equipment if the base is filled to 85% of its capacity.

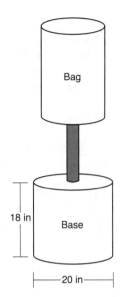

Part IV

Answer the question in this part. A correct answer will receive 6 credits. Clearly indicate the necessary steps, including appropriate formula substitutions, diagrams, graphs, charts, etc. Utilize the information provided for the question to determine your answer. Note that diagrams are not necessarily drawn to scale. A correct numerical answer with no work shown will receive only 1 credit. All answers should be written in pen, except for graphs and drawings, which should be done in pencil. [6]

35. Given: Parallelogram $ABCD$, $\overline{BF} \perp \overline{AFD}$, and $\overline{DE} \perp \overline{BEC}$

Prove: $BEDF$ is a rectangle

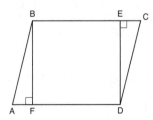

GEOMETRY
August 2018
Part I

Answer all **24** questions in this part. Each correct answer will receive **2 credits. No partial credit will be allowed.** Utilize the information provided for each question to determine your answer. Note that diagrams are not necessarily drawn to scale. For each statement or question, choose the word or expression that, of those given, best completes the statement or answers the question. Record your answers in the space provided [48]

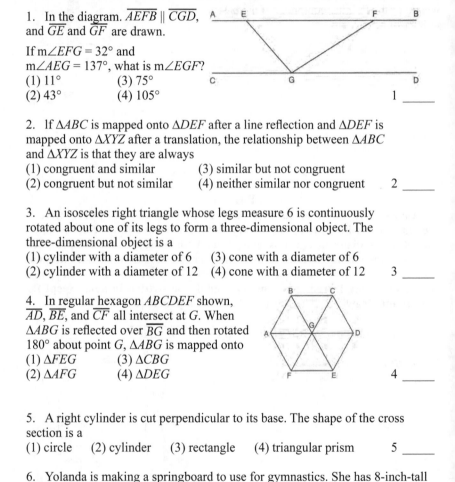

1. In the diagram. $\overline{AEFB} \parallel \overline{CGD}$, and $\overline{GE}$ and $\overline{GF}$ are drawn.

If m$\angle EFG = 32°$ and
m$\angle AEG = 137°$, what is m$\angle EGF$?
(1) 11°   (3) 75°
(2) 43°   (4) 105°   1 _____

2. If $\triangle ABC$ is mapped onto $\triangle DEF$ after a line reflection and $\triangle DEF$ is mapped onto $\triangle XYZ$ after a translation, the relationship between $\triangle ABC$ and $\triangle XYZ$ is that they are always
(1) congruent and similar   (3) similar but not congruent
(2) congruent but not similar   (4) neither similar nor congruent   2 _____

3. An isosceles right triangle whose legs measure 6 is continuously rotated about one of its legs to form a three-dimensional object. The three-dimensional object is a
(1) cylinder with a diameter of 6   (3) cone with a diameter of 6
(2) cylinder with a diameter of 12   (4) cone with a diameter of 12   3 _____

4. In regular hexagon $ABCDEF$ shown, $\overline{AD}$, $\overline{BE}$, and $\overline{CF}$ all intersect at G. When $\triangle ABG$ is reflected over $\overline{BG}$ and then rotated 180° about point G, $\triangle ABG$ is mapped onto
(1) $\triangle FEG$ (3) $\triangle CBG$
(2) $\triangle AFG$ (4) $\triangle DEG$ 4 _____

5. A right cylinder is cut perpendicular to its base. The shape of the cross section is a
(1) circle   (2) cylinder   (3) rectangle   (4) triangular prism   5 _____

6. Yolanda is making a springboard to use for gymnastics. She has 8-inch-tall springs and wants to form a 16.5° angle with the base, as modeled in the diagram below.

To the *nearest tenth of an inch*, what will be the length of the springboard, x?
(1) 2.3   (2) 8.3   (3) 27.0   (4) 28.2   6 _____

7. In the diagram of right triangle ABC, altitude $\overline{BD}$ is drawn to hypotenuse $\overline{AC}$.

If $BD = 4$, $AD = x - 6$, and $CD = x$, what is the length of $\overline{CD}$?

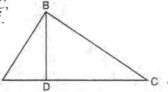

(1) 5   (3) 8

(2) 2   (4) 11   7 ____

8. Rhombus $STAR$ has vertices $S(-1, 2)$, $T(2, 3)$, $A(3, 0)$, and $R(0, -1)$. What is the perimeter of rhombus $STAR$?

(1) $\sqrt{34}$ (2) $4\sqrt{34}$ (3) $\sqrt{10}$ (4) $4\sqrt{10}$ 8 ____

9. In the diagram below of $\triangle HAR$ and $\triangle NTY$. angles H and N are right angles, and $\triangle HAR \sim \triangle NTY$.

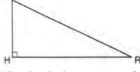
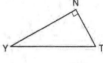

If $AR = 13$ and $HR = 12$, what is the measure of angle Y, to the *nearest degree*?

(1) $23°$ (2) $25°$ (3) $65°$ (4) $67°$ 9 ____

10. In the diagram, $\overline{AKS}$, $\overline{NKC}$, $\overline{AN}$, and $\overline{SC}$ are drawn such that $\overline{AN} \cong \overline{SC}$.

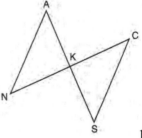

Which additional statement is sufficient to prove $\triangle KAN \cong \triangle KSC$ by AAS?

(1) $\overline{AS}$ and $\overline{NC}$ bisect each other.

(2) K is the midpoint of $\overline{NC}$.

(3) $\overline{AS} \perp \overline{CN}$

(4) $\overline{AN} \parallel \overline{SC}$ 10 ____

11. Which equation represents a line that is perpendicular to the line represented by $y = \frac{2}{3}x + 1$?

(1) $3x + 2y = 12$ (2) $3x - 2y = 12$ (3) $y = \frac{3}{2}x + 2$ (4) $y = -\frac{2}{3}x + 4$ 11 ____

12. in the diagram of $\triangle ABC$ below, points D and E are on sides $\overline{AB}$ and $\overline{CB}$ respectively, such that $\overline{DE} \parallel \overline{AC}$.

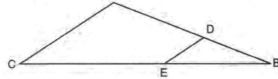

If EB is 3 more than DB, $AB = 14$, and $CB = 21$, what is the length of $\overline{AD}$?

(1) 6   (2) 8   (3) 9   (4) 12   12 ____

13. Quadrilateral *MATH* has both pairs of opposite sides congruent and parallel. Which statement about quadrilateral *MATH* is always true?
(1) $\overline{MT} \cong \overline{AH}$ (3) $\angle MHT \cong \angle ATH$
(2) $\overline{MT} \perp \overline{AH}$ (4) $\angle MAT \cong \angle MHT$ 13 ____

14. In the figure shown, quadrilateral *TAEO* is circumscribed around circle *D*. The midpoint of $\overline{TA}$ is *R*, and $\overline{HO} \cong \overline{PE}$.

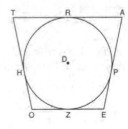

If *AP* = 10 and *EO* = 12, what is the perimeter of quadrilateral *TAEO*?
(1) 56   (3) 72
(2) 64   (4) 76   14 ____

15. The coordinates of the endpoints of directed line segment *ABC* are *A*(−8, 7) and *C*(7, −13). If *AB:BC* = 3:2, the coordinates of *B* are
(1) (1, −5)   (2) (−2, −1)   (3) (−3, 0)   (4) (3, −6)   15 ____

16. In triangle *ABC*, points *D* and *E* are on sides $\overline{AB}$ and $\overline{BC}$, respectively, such that $\overline{DE} \parallel \overline{AC}$, and *AD:DB* = 3:5.

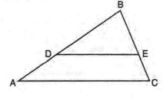

If *DB* = 6.3 and *AC* = 9.4, what is the length of $\overline{DE}$, to the *nearest tenth*?
(1) 3.8   (2) 5.6   (3) 5.9   (4) 15.7   16 ____

17. In the diagram, rectangle *ABCD* has vertices whose coordinates are *A*(7, 1), *B*(9, 3), *C*(3, 9), and *D*(1, 7).

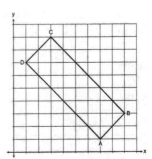

Which transformation will *not* carry the rectangle onto itself?
(1) a reflection over the line *y* = *x*
(2) a reflection over the line *y* = −*x* + 10
(3) a rotation of 180° about the point (6, 6)
(4) a rotation of 180° about the point (5, 5)   17 ____

18. A circle with a diameter of 10 cm and a central angle of 30° is drawn. What is the area, to the *nearest tenth of a square centimeter*, of the sector formed by the 30° angle?
(1) 5.2   (3) 13.1
(2) 6.5   (4) 26.2

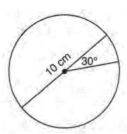

18 ____

August 2018

19. A child's tent can be modeled as a pyramid with a square base whose sides measure 60 inches and whose height measures 84 inches. What is the volume of the tent, to the *nearest cubic foot*?

(1) 35   (2) 58   (3) 82   (4) 175   19 _____

20. In the accompanying diagram of right triangle ABC, altitude $\overline{BD}$ is drawn to hypotenuse $\overline{AC}$.

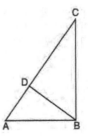

Which statement must always be true?

(1) $\dfrac{AD}{AB} = \dfrac{BC}{AC}$ (3) $\dfrac{BD}{BC} = \dfrac{AB}{AD}$

(2) $\dfrac{AD}{AB} = \dfrac{AB}{AC}$ (4) $\dfrac{AB}{BC} = \dfrac{BD}{AC}$ 20 _____

21. An equation of circle O is $x^2 + y^2 + 4x - 8y = -16$. The statement that best describes circle O is the

(1) center is $(2, -4)$ and is tangent to the x-axis
(2) center is $(2, -4)$ and is tangent to the y-axis
(3) center is $(-2, 4)$ and is tangent to the x-axis
(4) center is $(-2, 4)$ and is tangent to the y-axis   21 _____

22. In $\triangle ABC$, $\overline{BD}$ is the perpendicular bisector of $\overline{ADC}$. Based upon this information, which statements below can be proven?

         I. $\overline{BD}$ is a median.
         II. $\overline{BD}$ bisects $\angle ABC$.
         III. $\triangle ABC$ is isosceles.

(1) I and II, only   (3) II and III, only
(2) I and III, only   (4) I, II, and III   22 _____

23. Triangle RJM has an area of 6 and a perimeter of 12. If the triangle is dilated by a scale factor of 3 centered at the origin, what are the area and perimeter of its image, triangle $R'J'M'$?

(1) area of 9 and perimeter of 15
(2) area of 18 and perimeter of 36
(3) area of 54 and perimeter of 36
(4) area of 54 and perimeter of 108   23 _____

24. If $\sin (2x + 7)° = \cos (4x - 7)°$, what is the value of x?

(1) 7   (2) 15   (3) 21   (4) 30   24 _____

GEOMETRY
August 2018
Part II

Answer all **7** questions in this part. Each correct answer will receive **2 credits. Clearly indicate the necessary steps, including appropriate formula substitutions, diagrams, graphs, charts, etc. Utilize the information provided for each question to determine your answer. Note that diagrams are not necessarily drawn to scale. For all questions in this part, a correct numerical answer with no work shown will receive only 1 credit. All answers should be written in pen, except for graphs and drawings, which should be done in pencil.** [14]

25. In the circle below, $\overline{AB}$ is a chord. Using a compass and straightedge, construct a diameter of the circle. [Leave all construction marks.]

26. In parallelogram *ABCD* shown, the bisectors of $\angle ABC$ and $\angle DCB$ meet at *E*, a point on $\overline{AD}$.

If m$\angle A = 68°$, determine and state m$\angle BEC$.

27. In circle *A*, chord $\overline{BC}$ and diameter $\overline{DAE}$ intersect at *F*.

If m$\overarc{CD} = 46°$ and m$\overarc{DB} = 102°$, what is m$\angle CFE$?

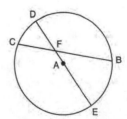

28. Trapezoids *ABCD* and *A"B"C"D"* are graphed on the set of axes.

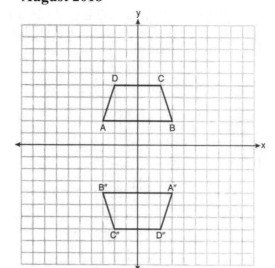

Describe a sequence of transformations that maps trapezoid *ABCD* onto trapezoid *A"B"C"D"*.

29. In the model, a support wire for a telephone pole is attached to the pole and anchored to a stake in the ground 15 feet from the base of the telephone pole. Jamal places a 6-foot wooden pole under the support wire parallel to the telephone pole, such that one end of the pole is on the ground and the top of the pole is touching the support wire. He measures the distance between the bottom of the pole and the stake in the ground.

Jamal says he can approximate how high the support wire attaches to the telephone pole by using similar triangles. Explain why the triangles are similar.

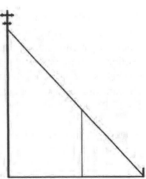

30. Aliyah says that when the line $4x + 3y = 24$ is dilated by a scale factor of 2 centered at the point $(3, 4)$, the equation of the dilated line is $y = -\frac{4}{3}x + 16$. Is Aliyah correct? Explain why. [The use of the set of axes below is optional.]

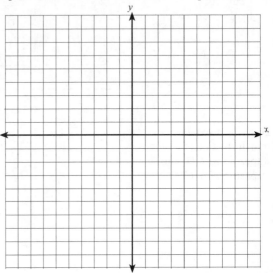

31. Ian needs to replace two concrete sections in his sidewalk, as modeled. Each section is 36 inches by 36 inches and 4 inches deep. He can mix his own concrete for $3.25 per cubic foot.

How much money will it cost Ian to replace the two concrete sections?

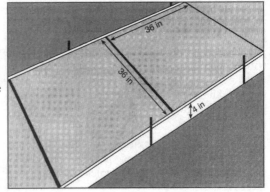

Answer all 3 questions in this part. Each correct answer will receive 4 credits. Clearly indicate the necessary steps, including appropriate formula substitutions, diagrams, graphs, charts, etc. Utilize the information provided for each question to determine your answer. Note that diagrams are not necessarily drawn to scale. For all questions in this part, a correct numerical answer with no work shown will receive only 1 credit. All answers should be written in pen, except for graphs and drawings, which should be done in pencil. [12]

32. Given: $\triangle ABC$, $\overline{AEC}$, $\overline{BDE}$ with $\angle ABE \cong \angle CBE$, and $\angle ADE \cong \angle CDE$

Prove: $\overline{BDE}$ is the perpendicular bisector of $\overline{AC}$.

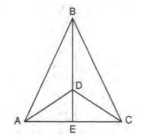

Fill in the missing statement and reasons below.

Statements	Reasons
(1) $\triangle ABC$, $\overline{AEC}$, $\overline{BDE}$ with $\angle ABE \cong \angle CBE$, and $\angle ADE \cong \angle CDE$	(1) Given
(2) $\overline{BD} \cong \overline{BD}$	(2) _____
(3) $\angle BDA$ and $\angle ADE$ are supplementary. $\angle BDC$ and $\angle CDE$ are supplementary.	(3) Linear pairs of angles are supplementary.
(4) _____	(4) Supplements of congruent angles are congruent
(5) $\triangle ABD \cong \triangle CBD$	(5) ASA
(6) $\overline{AD} \cong \overline{CD}$, $\overline{AB} \cong \overline{CB}$	(6) _____
(7) $\overline{BDE}$ is the perpendicular bisector of $\overline{AC}$.	(7) _____

GEOMETRY
August 2018

33. A homeowner is building three steps leading to a deck, as modeled by the diagram below. All three step rises, $\overline{HA}$, $\overline{FG}$, and $\overline{DE}$, are congruent, and all three step runs, $\overline{HG}$, $\overline{FE}$, and $\overline{DC}$, are congruent. Each step rise is perpendicular to the step run it joins. The measure of $\angle CAB = 36°$ and m$\angle CBA = 90°$.

If each step run is parallel to $\overline{AB}$ and has a length of 10 inches, determine and state the length of each step rise, to the *nearest tenth of an inch*.

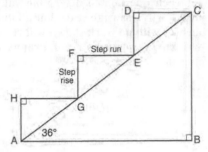

Determine and state the length of $\overline{AC}$, to the *nearest inch*.

34. A bakery sells hollow chocolate spheres. The larger diameter of each sphere is 4 cm. The thickness of the chocolate of each sphere is 0.5 cm. Determine and state, to the *nearest tenth of a cubic centimeter*, the amount of chocolate in each hollow sphere.

The bakery packages 8 of them into a box. If the density of the chocolate is 1.308 g/cm³, determine and state, to the *nearest gram*, the total mass of the chocolate in the box.

Answer the question in this part. A correct answer will receive 6 credits. Clearly indicate the necessary steps, including appropriate formula substitutions, diagrams, graphs, charts, etc. Utilize the information provided for the question to determine your answer. Note that diagrams are not necessarily drawn to scale. A correct numerical answer with no work shown will receive only 1 credit. All answers should be written in pen, except for graphs and drawings, which should be done in pencil. [6]

35. The vertices of quadrilateral *MATH* have coordinates $M(-4, 2)$, $A(-1, -3)$, $T(9, 3)$, and $H(6, 8)$.

Prove that quadrilateral *MATH* is a parallelogram.
[The use of the set of axes below is optional.]

Prove that quadrilateral *MATH* is a rectangle.
[The use of the set of axes below is optional.]

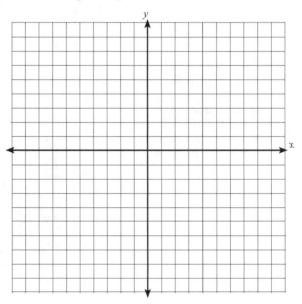

GEOMETRY
January 2019
Part I

Answer all 24 questions in this part. Each correct answer will receive 2 credits. No partial credit will be allowed. Utilize the information provided for each question to determine your answer. Note that diagrams are not necessarily drawn to scale. For each statement or question, choose the word or expression that, of those given, best completes the statement or answers the question. Record your answers in the space provided [48]

1. After a dilation with center $(0, 0)$, the image of $\overline{DB}$ is $\overline{D'B'}$. If $DB = 4.5$ and $D'B' = 18$, the scale factor of this dilation is

(1) $\frac{1}{5}$ (2) 5   (3) $\frac{1}{4}$ (4) 4   1 _____

2. In the diagram, $\triangle ABC$ with sides of 13, 15, and 16, is mapped onto $\triangle DEF$ after a clockwise rotation of 90° about point P.

If $DE = 2x - 1$, what is the value of x?

(1) 7   (2) 7.5   (3) 8   (4) 8.5   2 _____

3. On the set of axes, $\triangle ABC$ has vertices at $A(-2, 0)$, $B(2, -4)$, $C(4, 2)$, and $\triangle DEF$ has vertices at $D(4, 0)$, $E(-4, 8)$, $F(-8, -4)$.

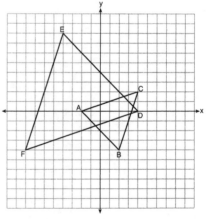

Which sequence of transformations will map $\triangle ABC$ onto $\triangle DEF$?

(1) a dilation of $\triangle ABC$ by a scale factor of 2 centered at point A

(2) a dilation of $\triangle ABC$ by a scale factor of $\frac{1}{2}$ centered at point A

(3) a dilation of $\triangle ABC$ by a scale factor of 2 centered at the origin, followed by a rotation of 180° about the origin

(4) a dilation of $\triangle ABC$ by a scale factor of $\frac{1}{2}$ centered at the origin, followed by a rotation of 180° about the origin   3 _____

4. The figure shows a rhombus with noncongruent diagonals. Which transformation would *not* carry this rhombus onto itself?

(1) a reflection over the shorter diagonal
(2) a reflection over the longer diagonal
(3) a clockwise rotation of 90° about the intersection of the diagonals
(4) a counterclockwise rotation of 180° about the intersection of the diagonals

4 _____

5. In the diagram of circle O, points K, A, T, I, and E are on the circle, $\triangle KAE$ and $\triangle ITE$ are drawn, $\overline{KE} \cong \overline{EI}$, and $\angle EKA \cong \angle EIT$. Which statement about $\triangle KAE$ and $\triangle ITE$ is always true?

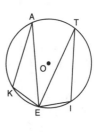

(1) They are neither congruent nor similar.
(2) They are similar but not congruent.
(3) They are right triangles.
(4) They are congruent.

5 _____

6. In right triangle ABC shown, point D is on $\overline{AB}$ and point E is on $\overline{CB}$ such that $\overline{AC} \| \overline{DE}$.

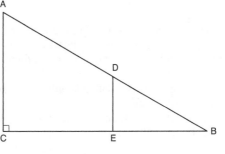

If $AB = 15$, $BC = 12$, and $EC = 7$, what is the length of $\overline{BD}$?

(1) 8.75   (2) 6.25   (3) 5   (4) 4   6 _____

7. In rhombus $VENU$, diagonals $\overline{VN}$ and $\overline{EU}$ intersect at S. If $VN = 12$ and $EU = 16$, what is the perimeter of the rhombus?
(1) 80   (2) 40   (3) 20   (4) 10   7 _____

8. Given right triangle ABC with a right angle at C, $m\angle B = 61°$.
Given right triangle RST with a right angle at T, $m\angle R = 29°$.

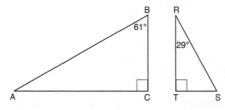

Which proportion in relation to $\triangle ABC$ and $\triangle RST$ is *not* correct?

(1) $\dfrac{AB}{RS} = \dfrac{RT}{AC}$ (2) $\dfrac{BC}{ST} = \dfrac{AB}{RS}$ (3) $\dfrac{BC}{ST} = \dfrac{AC}{RT}$ (4) $\dfrac{AB}{AC} = \dfrac{RS}{RT}$ 8 _____

9. A vendor is using an 8-ft by 8-ft tent for a craft fair. The legs of the tent are 9 ft tall and the top forms a square pyramid with a height of 3 ft.

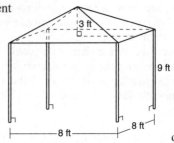

What is the volume, in cubic feet, of space the tent occupies?
(1) 256   (3) 672
(2) 640   (4) 768

9 _____

10. In the diagram of right triangle *KMI*, altitude $\overline{IG}$ is drawn to hypotenuse $\overline{KM}$. If $KG = 9$ and $IG = 12$, the length of $\overline{IM}$ is
(1) 15   (3) 20
(2) 16   (4) 25

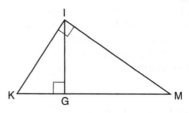

10 _____

11. Which three-dimensional figure will result when a rectangle 6 inches long and 5 inches wide is continuously rotated about the longer side?
(1) a rectangular prism with a length of 6 inches, width of 6 inches, and height of 5 inches
(2) a rectangular prism with a length of 6 inches, width of 5 inches, and height of 5 inches
(3) a cylinder with a radius of 5 inches and a height of 6 inches
(4) a cylinder with a radius of 6 inches and a height of 5 inches

11 _____

12. Which statement about parallelograms is always true?
(1) The diagonals are congruent.
(2) The diagonals bisect each other.
(3) The diagonals are perpendicular.
(4) The diagonals bisect their respective angles.

12 _____

13. From a point on the ground one-half mile from the base of a historic monument, the angle of elevation to its top is 11.87°. To the *nearest foot*, what is the height of the monument?
(1) 543   (2) 555   (3) 1086   (4) 1110

13 _____

14. The area of a sector of a circle with a radius measuring 15 cm is 75π cm². What is the measure of the central angle that forms the sector?
(1) 72°   (2) 120°   (3) 144°   (4) 180°

14 _____

15. Point *M* divides $\overline{AB}$ so that *AM:MB* = 1:2. If *A* has coordinates (−1, −3) and *B* has coordinates (8, 9), the coordinates of *M* are

(1) (2, 1)   (2) $(\frac{5}{3}, 0)$ (3) (5, 5)   (4) $(\frac{23}{3}, 8)$

15 _____

16. In the diagram of triangle ABC, $\overline{AC}$ is extended through point C to point D, and $\overline{BE}$ is drawn to $\overline{AC}$. Which equation is always true?

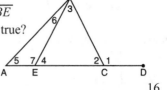

(1) $m\angle 1 = m\angle 3 + m\angle 2$
(2) $m\angle 5 = m\angle 3 - m\angle 2$
(3) $m\angle 6 = m\angle 3 - m\angle 2$
(4) $m\angle 7 = m\angle 3 + m\angle 2$

16 ____

17. In the diagram of right triangle ABC, $AC = 8$, and $AB = 17$.

Which equation would determine the value of angle A?

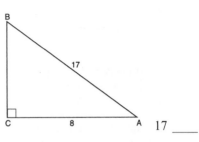

(1) $\sin A = \dfrac{8}{17}$ (3) $\cos A = \dfrac{15}{17}$

(2) $\tan A = \dfrac{8}{15}$ (4) $\tan A = \dfrac{15}{8}$

17 ____

18. Francisco needs the three pieces of glass shown below to complete a stained glass window. The shapes, two triangles and a trapezoid, are measured in inches.

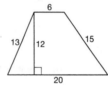

Glass can be purchased in rectangular sheets that are 12 inches wide. What is the minimum length of a sheet of glass, in inches, that Francisco must purchase in order to have enough to complete the window?

(1) 20   (2) 25   (3) 29   (4) 34   18 ____

19. In the diagram of quadrilateral $NAVY$ below, $m\angle YNA = 30°$, $m\angle YAN = 38°$, $m\angle AVY = 94°$, and $m\angle VAY = 46°$.

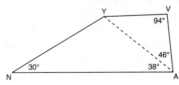

Which segment has the shortest length?

(1) $\overline{AY}$ (2) $\overline{NY}$ (3) $\overline{VA}$ (4) $\overline{VY}$ 19 ____

20. What is an equation of a circle whose center is $(1, 4)$ and diameter is 10?

(1) $x^2 - 2x + y^2 - 8y = 8$ (3) $x^2 - 2x + y^2 - 8y = 83$
(2) $x^2 + 2x + y^2 + 8y = 8$ (4) $x^2 + 2x + y^2 + 8y = 83$ 20 ____

21. On the set of axes, $\triangle ABC$, altitude $\overline{CG}$, and median $\overline{CM}$ are drawn.

Which expression represents the area of $\triangle ABC$?

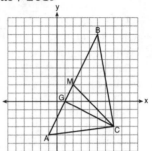

(1) $\dfrac{(BC)(AC)}{2}$ (3) $\dfrac{(CM)(AB)}{2}$

(2) $\dfrac{(GC)(BC)}{2}$ (4) $\dfrac{(GC)(AB)}{2}$

21 ____

22. In right triangle ABC, $m\angle C = 90°$ and $AC \neq BC$. Which trigonometric ratio is equivalent to sin B?

(1) cos A (2) cos B (3) tan A (4) tan B 22 ____

23. As shown in the diagram, the radius of a cone is 2.5 cm and its slant height is 6.5cm. How many cubic centimeters are in the volume of the cone?

(1) 12.5π

(2) 13.5π

(3) 30.0π

(4) 37.5π 23 ____

24. What is an equation of the image of the line $y = \dfrac{3}{2}x - 4$ after a dilation of a scale factor of $\dfrac{3}{4}$ centered at the origin?

(1) $y = \dfrac{9}{8}x - 4$ (2) $y = \dfrac{9}{8}x - 3$ (3) $y = \dfrac{3}{2}x - 4$ (4) $y = \dfrac{3}{2}x - 3$ 24 ____

Part II

Answer all 7 questions in this part. Each correct answer will receive 2 credits. Clearly indicate the necessary steps, including appropriate formula substitutions, diagrams, graphs, charts, etc. Utilize the information provided for each question to determine your answer. Note that diagrams are not necessarily drawn to scale. For all questions in this part, a correct numerical answer with no work shown will receive only 1 credit. All answers should be written in pen, except for graphs and drawings, which should be done in pencil. [14]

25. Write an equation of the line that is parallel to the line whose equation is $3y + 7 = 2x$ and passes through the point $(2, 6)$.

26. Parallelogram *ABCD* is adjacent to rhombus *DEFG*, as shown, and $\overline{FC}$ intersects $\overline{AGD}$ at *H*.

If m∠*B* = 118° and m∠*AHC* = 138°, determine and state m∠*GFH*.

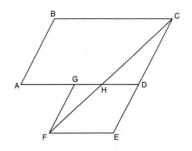

27. As shown in the diagram, secants $\overrightarrow{PWR}$ and $\overrightarrow{PTS}$ are drawn to circle *O* from external point *P*.

If m∠*RPS* = 35°and m$\overset{\frown}{RS}$ = 121°, determine and state m$\overset{\frown}{WT}$.

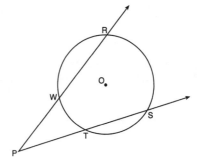

January 2019

28. On the set of axes, △ABC is graphed with coordinates A(−2, −1), B(3, −1), and C(−2, −4). Triangle QRS, the image of △ABC, is graphed with coordinates Q(−5, 2), R(−5, 7), and S(−8, 2).

Describe a sequence of transformations that would map △ABC onto △QRS.

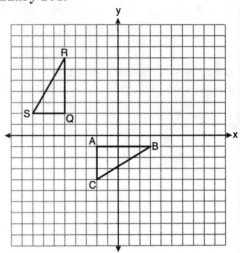

29. Given points A, B, and C, use a compass and straightedge to construct point D so that ABCD is a parallelogram. [Leave all construction marks.]

•C

•A   •B

30. On the set of axes, $\triangle DEF$ has
vertices at the coordinates $D(1, -1)$,
$E(3, 4)$, and $F(4, 2)$, and point G
has coordinates $(3, 1)$. Owen claims
the median from point E must pass
through point G.

Is Owen correct? Explain why.

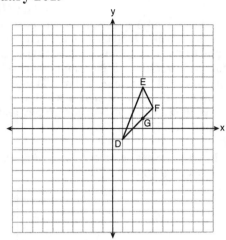

31. A walking path at a local park is modeled on the grid below, where the
length of each grid square is 10 feet. The town needs to submit paperwork to
pave the walking path. Determine and state, to the *nearest square foot*, the area
of the walking path.

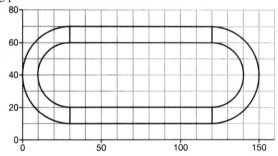

GEOMETRY
January 2019
Part III

Answer all 3 questions in this part. Each correct answer will receive 4 credits. Clearly indicate the necessary steps, including appropriate formula substitutions, diagrams, graphs, charts, etc. Utilize the information provided for each question to determine your answer. Note that diagrams are not necessarily drawn to scale. For all questions in this part, a correct numerical answer with no work shown will receive only 1 credit. All answers should be written in pen, except for graphs and drawings, which should be done in pencil. [12]

32. A triangle has vertices $A(-2, 4)$, $B(6, 2)$, and $C(1, -1)$. Prove that $\triangle ABC$ is an isosceles right triangle. [The use of the set of axes below is optional.]

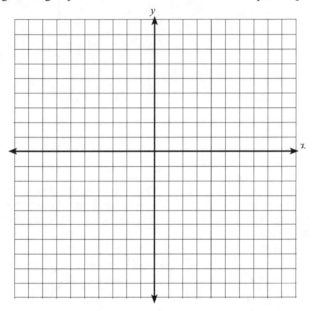

33. Theresa has a rectangular pool 30 ft long, 15 ft wide, and 4 ft deep. Theresa fills her pool using city water at a rate of $3.95 per 100 gallons of water.

Nancy has a circular pool with a diameter of 24 ft and a depth of 4 ft. Nancy fills her pool with a water delivery service at a rate of $200 per 6000 gallons.

If Theresa and Nancy both fill their pools 6 inches from the top of the pool, determine and state who paid more to fill her pool. [1 ft³ water = 7.48 gallons]

34. As modeled in the diagram below, an access ramp starts on flat ground and ends at the beginning of the top step. Each step is 6 inches tall and 8 inches deep.

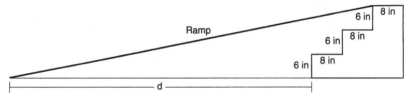

If the angle of elevation of the ramp is 4.76°, determine and state the length of the ramp, to the *nearest tenth of a foot*.

Determine and state, to the *nearest tenth of a foot*, the horizontal distance, *d*, from the bottom of the stairs to the bottom of the ramp.

GEOMETRY
January 2019
Part IV

Answer the question in this part. A correct answer will receive 6 credits. Clearly indicate the necessary steps, including appropriate formula substitutions, diagrams, graphs, charts, etc. Utilize the information provided for each question to determine your answer. Note that diagrams are not necessarily drawn to scale. A correct numerical answer with no work shown will receive only 1 credit. All answers should be written in pen, except for graphs and drawings, which should be done in pencil. [6]

35. In the diagram of quadrilateral $ABCD$ with diagonal $\overline{AC}$ shown, segments GH and EF are drawn, $\overline{AE} \cong \overline{CG}$, $\overline{BE} \cong \overline{DG}$, $\overline{AH} \cong \overline{CF}$, and $\overline{AD} \cong \overline{CB}$.

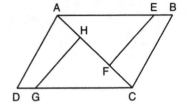

Prove: $\overline{EF} \cong \overline{GH}$